Human Body

Written by Angela Royston
Illustrated by Mike Saunders

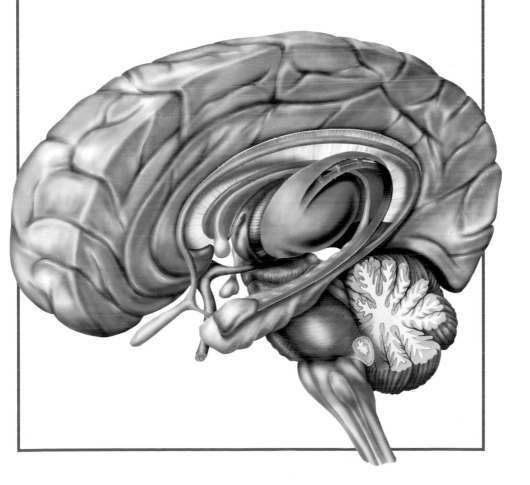

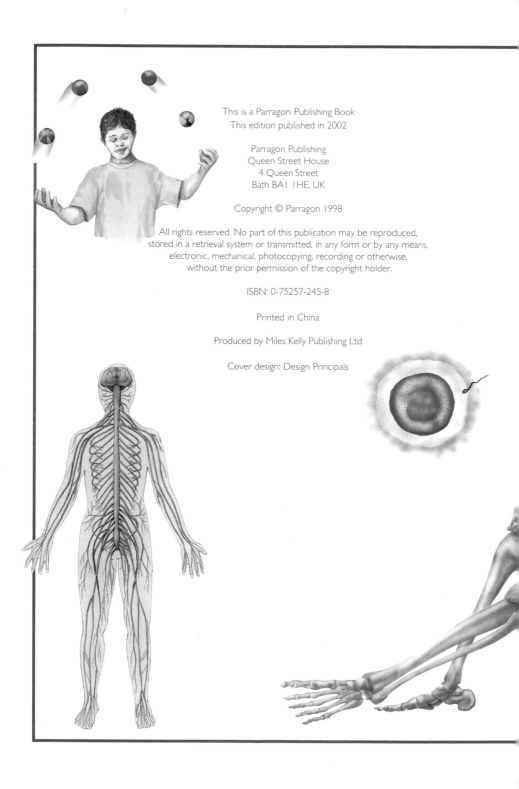

This is a Parragon Publishing Book
This edition published in 2002

Parragon Publishing
Queen Street House
4 Queen Street
Bath BA1 1HE, UK

ISBN: 0-75257-245-8

Printed in China

Produced by Miles Kelly Publishing Ltd

Cover design: Design Principals

Contents

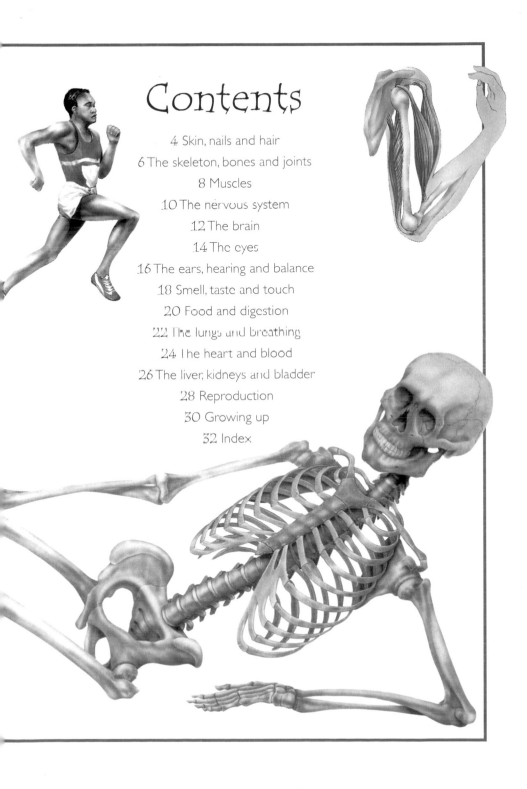

A strand of hair

Most people have about 100,000 hairs growing on their head.

What gives hair its color?

The color of your hair is determined mainly by the pigments (colored substances) it contains. There are two kinds of pigments – melanin, which is very dark brown, and carotene, which is reddish yellow – and all hair color is formed by one or other or a mixture of both.

Why do old people have gray hair?

Some people's hair stops making melanin as they grow older. Fair-haired people tend to go white, while dark-haired people usually go gray. Grayness is lack of melanin plus tiny air bubbles in the hair.

Why does skin have pores?

Skin has tiny holes in it called sweat pores to let out sweat and water vapor. When you are too hot, glands pump out sweat, which cools you as it dries.

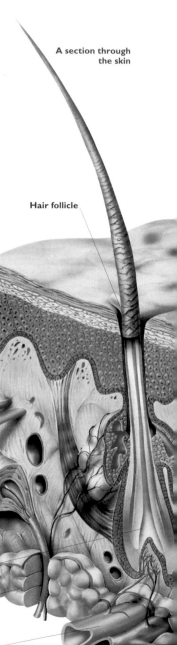

A section through the skin

Hair follicle

Melanin

Dermis

Hair root

Fat

Blood vessel

How thick is your skin?

S KIN VARIES IN THICKNESS FROM ABOUT $1/25$ IN (1 MM) ON YOUR EYELIDS to $1/5$ in (5 mm) or more on the soles of your feet. This thin covering has two layers. The outer skin that you see and touch is made of hard, tough cells of dead skin and is part of the epidermis. Below the epidermis is the dermis. It contains tiny blood vessels, sweat glands, nerve endings, and the roots of tiny hairs. Under the dermis is a layer of fat, which keeps you warm.

What are goose bumps?

Goose bumps are bumps on your skin formed by the tiny muscles that make the hairs on your skin stand up when you are cold.

What are freckles?

Freckles are small patches of darker skin made by extra melanin. Exposure to sunshine increases the amount of melanin in your skin and the darkness of freckles.

What makes fingerprints unique?
A fingerprint is made by thin ridges of skin on the tip of each finger and thumb. The ridges form a pattern of lines, loops, or whorls and no two people have the same pattern.

How do nails grow, and how fast do they grow?
Each nail grows about $1/25$ in (1mm) every 10 days. As new nail forms behind the cuticle, under the skin, it pushes the older nail along.

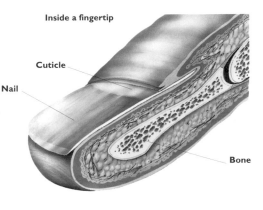

Inside a fingertip

Cuticle

Nail

Bone

Epidermis

Sweat pore

What makes hair naturally curly?
How curly your hair is depends on the shape of the follicle, the tiny pit from which it grows. Curly hair grows from flat hair follicles, wavy hair from oval follicles, and straight hair from round follicles.

Why does hair fall out?
No hair lasts more than about six years. Every day you lose about 60 hairs, but since you have about 100,000 on your scalp you hardly notice. After a while new hairs grow from the hair follicles.

What does skin do?

SKIN IS A TOUGH, STRETCHY COVERING THAT ACTS AS a barrier between your body and the outside world. It stops the moisture inside the body from drying out and it prevents dirt and germs from getting in. Tiny particles of melanin inside the epidermis help to shield your body from the harmful rays of the Sun. The more melanin you have, the darker your skin is and the better protected you are.

Sweat gland

Which is the longest bone?
The thigh bone in the upper part of your leg is the longest bone in your body. It accounts for more than a quarter of an adult's height.

Which is the smallest bone?
The smallest bone is called the stirrup and is no bigger than a grain of rice. It is deep inside your ear and its job is to pass on sounds from the outer and middle ear to the inner ear.

Why do bodies need bones?

BONES PROVIDE A STRONG FRAMEWORK THAT SUPPORTS THE REST OF THE BODY. Without bones, you would flop on the floor like an octopus. Some of the bones form a suit of internal armor, which protects the brain, the lungs, the heart, and other vital organs. All the bones together are called the skeleton. You can move and bend different parts of the body because the bones meet at joints.

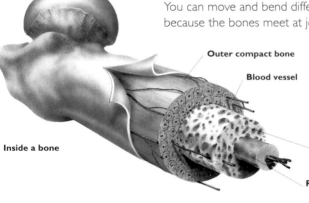

Outer compact bone

Blood vessel

Inside a bone

What's inside a bone?
Inside a bone is a crisscross honeycomb of lighter bone. Blood vessels weave in and out of the bone, keeping the cells alive. At the centre of some bones is a core of bone marrow.

Softer spongy bone

Red marrow jelly

How many bones do you have?
As a baby you had over 300 bones, but, as you grow, some bones join together. When you are an adult, you will have about 206 bones in total.

Which joints move the least?
Your skull is made up of more than 20 bones fused together in joints that allow no movement at all. These are called suture joints.

What are ligaments?
Ligaments are strong, bendy straps that hold together the bones in a joint. Nearly all the body's joints have several ligaments.

What is a vertebra?
A vertebra is a knobbly bone in your spine. The 26 vertebrae fit together to make a strong pillar, the spine, which carries much of your weight. At the same time, the vertebrae allow your back to bend and twist.

Which joint moves the most?
The shoulder joint is a ball-and-socket joint and it allows the greatest amount of movement in all directions.

The human skeleton

Finger bones

Patella (kneecap)

Tibia (shin)

Fibula

Femur (thigh)

Toe bones

Calcaneus (heel bone)

Ball-and-socket joint

The shoulder and hip have this joint.

Pivot joint

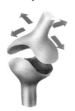

The pivot joint is found in the neck.

Saddle joint

This joint is found at the base of the thumb.

Hinge joint

A joint that works like a hinge is found at the knee and elbow, and in the fingers and toes.

What is a joint?

WHERE TWO BONES MEET, THEIR ENDS ARE SHAPED TO MAKE DIFFERENT kinds of joints. Each kind of joint makes a strong connection and allows a particular kind of movement. For example, the knee is a hinge joint that lets the lower leg move only back and forward. The hip is a ball-and-socket joint that allows you to move your thigh around in a circle. The saddle joint at the base of the thumb also gives a good range of movement.

Why do joints not squeak?
Joints are cushioned by soft, squashy cartilage. Many joints also contain a fluid – called synovial fluid – that works like oil to keep them moving smoothly and painlessly.

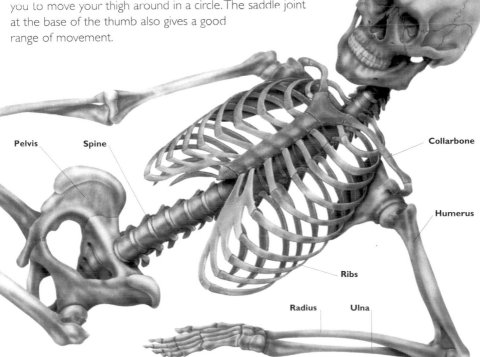

Skull

Pelvis

Spine

Collarbone

Humerus

Ribs

Radius

Ulna

7

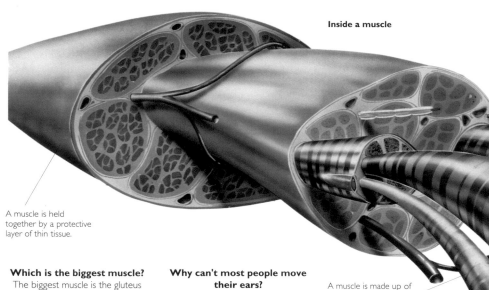

Inside a muscle

A muscle is held together by a protective layer of thin tissue.

A muscle is made up of many bundles of fibers.

Which is the biggest muscle?
The biggest muscle is the gluteus maximus in the buttock. You use it to straighten your leg when you stand up and it makes a comfortable cushion when you sit down.

Why can't most people move their ears?
Humans, like most other animals, have a muscle behind each ear. Animals can turn their ears to hear better, but most people never learn how to use their ear muscles.

How do muscles work?
Muscles work by contracting. This makes them shorter and thicker so that they pull on whatever bone or other part of the body they are attached to, thereby making it move.

Why does exercise make muscles stronger?
A muscle is made of bundles of fibers that contract when you use the muscle. The more you use the muscle, the thicker the fibres become. They contract more effectively, which means the muscle is stronger.

The body's muscles

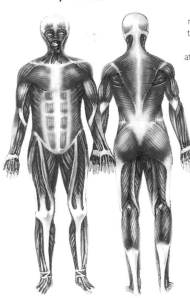

Here the skin is stripped away to show the muscles of the front and back of the body.

What do the muscles do?

MUSCLES MAKE PARTS OF YOUR BODY MOVE. THE SKELETON IS COVERED with muscles that move your bones and give your body its shape. Muscles in the face move your cheeks, eyebrows, nose, mouth, tongue, and lower jaw. A different kind of muscle works in the esophagus (food pipe), stomach, and intestines to move food through your body. The heart is a third type of muscle – it never stops beating to move blood around your body.

Why do muscles work in pairs?

MUSCLES CANNOT PUSH – THEY CAN ONLY PULL – AND SO YOU NEED TWO sets of muscles for many actions. For example, the biceps in your upper arm bends your elbow and you can feel it tighten when you clench your arm. To straighten the elbow again, you have to relax the biceps and tighten the triceps, which is the muscle at the back of your upper arm. In the same way, one set of muscles lifts the leg and another set of muscles straightens it.

Bending the arm

The biceps contracts to bend the elbow.

When the biceps contracts, the triceps is relaxed.

What is a tendon?

A tendon is like a tough rope that joins a muscle to a bone. If you bend and straighten your fingers, you can feel the tendons in the back of your hand. The body's strongest tendon is the Achilles tendon, which you can feel above your heel.

Each fiber is made up of hundreds of strands called fibrils.

How many muscles are there?

You have about 650 muscles that work together. Most actions – including walking, swimming, and smiling – involve dozens of muscles. Even frowning uses 40 different muscles, but smiling is less energetic – it uses only 15.

Muscle fibers are so small that 1/4 sq in (1 sq cm) would contain a million of them.

What do the nerves do?

Juggling

Juggling takes great skill. As the juggler learns to coordinate throwing and catching, the actions become automatic.

NERVES CARRY INFORMATION AND INSTRUCTIONS TO AND FROM THE brain and from one part of the brain to the other. Sensory nerves bring information from the eyes, ears, and other sense organs to the brain, and motor nerves control the muscles. For example, if you decide to bend your knee, electrical signals move from your brain to the muscles in your leg to make them contract.

What is a reflex action?

A reflex action is something you do automatically, without thinking about it. Swallowing, blinking, and choking are reflex actions. So is snatching your hand away from a hot plate.

How fast do nerves act?

A nerve signal is a tiny pulse of electricity. It travels at about 3 ft (1 meter) per second in the slowest nerves to more than 300 ft (100 meters) per second in the fastest ones.

Dendrites (the arms of the nerve cell) collect signals from other cells.

The cell body contains the nucleus of the cell.

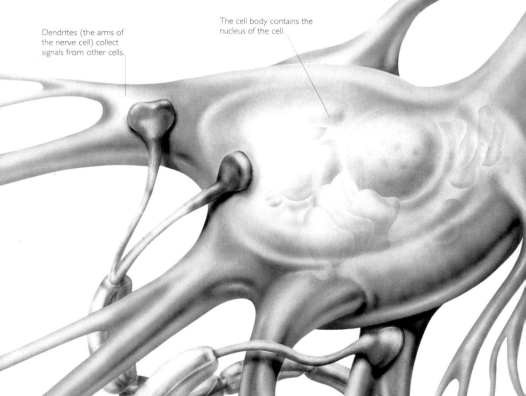

A single nerve cell

How does a nerve work?
A chain of nerve cells carries a signal to or from the brain. The electrical impulse is received by the nerve endings and sent through the first nerve cell and along its nerve fiber to the nerve endings of the next nerve cell.

What is the spinal cord?
The spinal cord is the largest nerve in the body. It is ³/₄ in (2 cm) wide and runs through the center of the spine. It connects the nerves in the body with the brain.

How do anesthetics work?
An anesthetic stops you feeling. A local anesthetic deadens the sensory nerves so that part of your body goes numb. A general anesthetic puts you into a deep sleep so that none of your senses are taking in information.

The axon is covered with a fatty myelin sheath, which acts as an insulator to stop the signals leaking away.

The axon carries nerve signals to the next nerve cell.

How many nerves do you have?
Billions of nerves reach out to all parts of the body.

Which is the longest nerve?
The longest nerve is the tibial nerve. It runs alongside the tibia (the shinbone), and in adults it is 20 in (50 cm) long.

What causes "pins and needles"?
If a nerve gets squashed, it cannot carry nerve signals. If you kneel for a long time, your leg goes numb and then, when you stretch it, it tingles as the signals begin to flow again.

The nervous system

Nerves link the brain and spinal cord to all parts of the body.

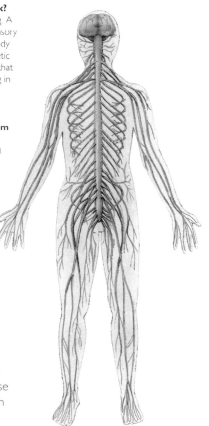

What are the body's five main senses?

THE FIVE MAIN SENSES ARE SEEING, HEARING, SMELLING, TASTING, AND touching. Between them the five senses give you all the information you have about the outside world. Each sense has a special part of the body, called a sense organ, which reacts to a particular kind of stimulus. For example, eyes react to light and ears react to sound.

What does the brain do?

YOUR BRAIN CONTROLS YOUR BODY. IT KEEPS THE HEART, STOMACH, LUNGS, kidneys, and other vital organs working. The information collected by the senses is processed by different parts of the brain. Some is discarded, some is stored, and some is reacted to at once, with messages being sent from the brain to the muscles and glands. The brain also gives you your sense of who you are. Memories of the past are stored here and everything you think, feel, and do is controlled by the brain.

What does the brain look like?
The brain looks soft and grayish pink. The top is wrinkled like a walnut and it is covered with many tiny tubes of blood. The spinal cord links the brain to the rest of the body.

What is the brain made of?
The brain consists of water and billions of nerve cells and nerve fibers. It is surrounded by protective coverings called the meninges.

Inside the brain

Skipping rope

When you skip rope, your brain coordinates balance with the movements of your arms and legs.

Why are some people more artistic than others?
One side of the brain deals more with music and artistic skills, and the other side deals more with logical skills. How artistic or mathematical you are depends on which side of your brain is dominant (stronger).

Why do you remember some things and forget others?
On the whole, you remember things that are important to you in some way. Some things need to be remembered for only a very short while. For instance, you might look up a telephone number, keep it in your head while you dial it, and then forget it completely.

The front of the cortex is mainly involved with thinking and planning.

The hypothalamus controls hunger, thirst, and body temperature.

The pituitary gland controls growth and many other body processes.

What does the skull do?
The skull is a hard covering of bone that protects the brain like a helmet. All the bones of the skull except the lower jaw are fused together to make them stronger.

Why are some people left-handed?

Most people are right-handed because the left side of their brain is dominant, but in left-handed people, the right side of the brain is dominant. The part of the brain that controls speech is usually on the dominant side.

The cerebral cortex covers most of the brain.

When you are asleep you are not aware of what is going on around you. The brain blocks incoming signals unless they are so strong they wake you up.

Why do you need to sleep?

A 10-year-old sleeps on average nine or 10 hours a night, but sleep time can vary a lot between four and 12 hours. If you sleep for eight hours a night, that's a third of your life! You need to sleep to rest your muscles and to allow your body time to repair and replace damaged cells.

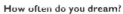

Sleeping

How often do you dream?

You probably dream about five times every night, but you are only aware of dreaming if you wake up during a dream.

Why do some people sleepwalk?

People may walk in their sleep because they are worried or anxious. If someone is sleepwalking you should gently take them back to bed.

Does the brain ever rest?

Even while you are asleep the brain carries on controlling body activities such as breathing, heartbeat, and digestion.

This part of the cortex deals with sight.

What does the cerebral cortex do?

THE CORTEX IS THE WRINKLY TOP PART OF THE BRAIN. IT CONTROLS ALL THE brain activity that you are aware of – seeing, thinking, reading, feeling, and moving. Only humans have such a large and well-developed cerebral cortex. Different parts of the cortex deal with different activities. The left side controls the right side of the body, while the right side of the cortex controls the left side of the body.

The cerebellum coordinates movement and balance.

How do the eyes see things?

YOU SEE SOMETHING WHEN LIGHT BOUNCES OFF IT AND ENTERS YOUR EYES. The black circle in the middle of the eye is a hole, called the pupil. Light passes through the pupil and is focused by the lens onto the retina at the back of the eye. Nerve endings in the retina send signals along the optic nerve to the brain. The picture formed on the retina is upside down, but the brain turns it around so that you perceive things to be the right way up.

Why does the pupil change size?
The pupil becomes smaller in bright light to stop too much light from getting in and damaging the retina. In dim light the pupil opens to let in more light. The iris is a muscle that opens and closes the pupil automatically.

The eyes

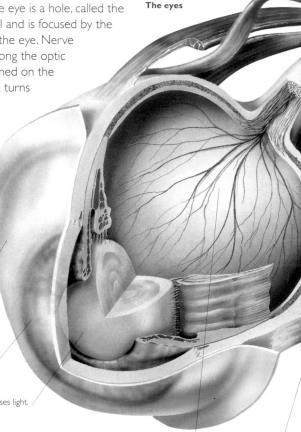

Why do people have different colored eyes?
The iris is the colored ring around the pupil. The color is formed by a substance called melanin – brown-eyed people have a lot of melanin, while blue-eyed people have very little.

How big is your eyeball?
An adult eyeball is about the size of a golf ball, but most of the eyeball is hidden inside your head.

The cornea is a tough, see-through layer that protects the eye.

The iris is a circular muscle that controls the size of the pupil to allow light into the eye.

The lens focuses light.

How light enters the eye

Light hits the retina at the back of the eye.

Muscles hold and move the eyeball.

The image of the object projected onto the retina at the back of the eye is upside down.

What is the blind spot?
The blind spot is a spot on the retina where the optic nerve leaves the eye. There are no light-sensitive cells here, making the spot "blind."

Can sunshine damage the eye?
Sunshine contains ultraviolet rays, which can damage your eyes as well as your skin. You should wear sunglasses in bright sunlight and never, never look directly at the Sun.

Why do you have two eyes?
Two eyes help you to judge how far away something is. Each eye gets a slightly different picture, which the brain combines into a single three-dimensional or 3-D picture. A 3-D picture is one that has depth as well as height and breadth.

What are eyelashes for?
Your eyelashes help to protect your eyes by stopping dust and dirt getting blown into them.

What keeps the eye in place?
The eyeball is held firmly in place by six muscles attached to the top, bottom, and each side of the eye. These muscles work together to move your eyes so that you can look around.

What makes you cry?
If dust or something gets into your eye, the tear gland above the eye releases extra tears to wash it away. Being upset can also make you cry.

The optic nerve takes signals from the retina to the brain.

The eyeball is filled with jelly, which keeps it in shape.

The tear gland makes a constant supply of salty water.

The pupil is the black hole at the center of the iris.

When too much water floods the eye, some spills over as tears and the rest drains into the nose.

The tear duct drains tears to the nose.

How do you see color?
Different nerve cells in the retina react to the colors red, blue, and green. Together they make up all the other colors.

Why can't you see color when it starts to get dark?
The cells that react to colored light – called cones – only work well in bright light. Most of the cells in the eye see in black, white, and gray, and these – called rods – are the ones that work at night.

Why do you blink?

YOU BLINK TO CLEAN AND PROTECT YOUR EYES. EACH EYE IS COVERED WITH A THIN film of salty water, so every time you blink, the eyelid washes the eyeball and wipes away dust and germs. The water drains away through a narrow tube into your nose. You also blink to protect your eye when something comes too close to it. Blinking is so important, you do it automatically.

How do you hear?

SOUND REACHES YOUR EARS AS VIBRATIONS IN THE AIR. The vibrations travel down the ear canal to the eardrum, which then vibrates, making the bones in the middle ear vibrate too. These three small bones make the vibrations bigger and pass them through to the liquid in the inner ear. The cochlea in the inner ear is coiled like a snail shell. As nerve endings in the lining of the cochlea detect vibrations in the liquid inside it, they send electrical signals to the brain.

The ear

Outer ear (pinna)

Why do you have two ears?
Two ears help you to detect which direction sounds are coming from.

How do you measure sound?
The loudness of a sound is measured in decibels. The sound of a pin dropping is less than 10 decibels, while the hum of a refrigerator is about 35 decibels. A loud personal stereo makes about 80 decibels, while the noise of a jet aircraft just 90 ft (30 meters) away can reach 130 decibels.

A jet lifts off the runway

The ear canal carries sound waves to the eardrum.

Why do your ears pop?
If you are flying in an aircraft and it changes height quickly, you may go a bit deaf, because the air inside and outside the eardrum are at different pressures. Your ears "pop" when the pressures become equal again.

The noise of a jet aircraft just 90 ft (30 meters) away can reach 130 decibels.

Is loud noise dangerous?
Any noise over about 120 decibels can damage your hearing immediately, but, if you are constantly listening to sounds of 90 decibels or more, they can damage your hearing too.

What is earwax?
This yellow-brown wax is made by glands in the skin lining the ear canal. Wax traps dirt and germs and is slowly pushed out of the ear.

Where does the Eustachian tube go?
This tube joins the middle ear to the empty spaces behind your upper throat. If mucus from a cold fills the tube, it stops you hearing as well as usual.

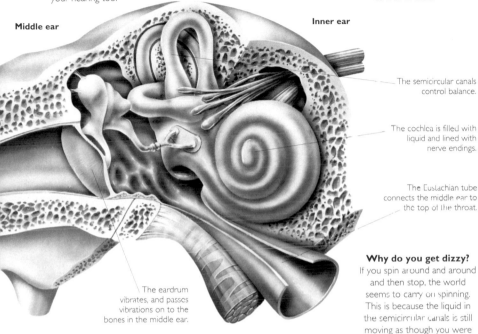

Middle ear

Inner ear

The semicircular canals control balance.

The cochlea is filled with liquid and lined with nerve endings.

The Eustachian tube connects the middle ear to the top of the throat.

The eardrum vibrates, and passes vibrations on to the bones in the middle ear.

Why do you get dizzy?
If you spin around and around and then stop, the world seems to carry on spinning. This is because the liquid in the semicircular canals is still moving as though you were still spinning.

How loud is a whisper?
A whisper is between 10 and 20 decibels. Some animals can detect much quieter sounds than we can.

What is sound?
Sound is waves of energy that are carried as vibrations through air, liquid, and solid objects.

Ballet dancer

A spinning dancer stops herself getting dizzy by turning her head quickly and keeping her eyes on just one thing.

How do ears help you balance?

THREE CURVED TUBES IN THE INNER EAR HELP YOU TO BALANCE. THEY ARE filled with liquid and are called the semicircular canals. They are arranged at right angles to each other (like three sides of a box) so that as you move, the liquid inside them moves too. Nerves in the lining of the tubes detect changes in the liquid and send the information to the brain.

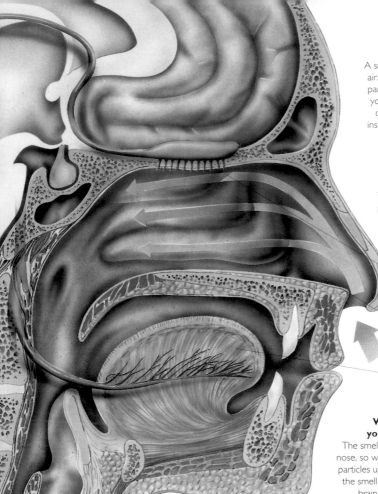

How do you smell?
A smell is made by tiny particles in the air. As you breathe in, some of these particles reach the smell receptors in your nose. Smell receptors react to chemicals dissolved in the mucus inside your nose and send a message to the brain.

Why do some things smell more than others?
Things that smell strongly, such as perfume or food cooking, give off more smell particles that float through the air.

Inside the nose and mouth

The smell receptors are situated at the top of the nose.

The inside of the nose is lined with mucus and fine hairs.

The tongue is a strong muscle covered with thousands of taste buds.

Why does sniffing help you detect smells better?
The smell receptors are at the top of your nose, so when you sniff you bring more smell particles up to them, which helps you detect the smell better. No one knows how your brain tells one smell from another.

Why does a blocked nose stop you tasting?
When you eat, you both taste and smell the food. If your nose is blocked with mucus from a cold, you can't smell properly and so food seems to have less taste too.

Why do some animals have keener smell?
Many animals rely on smell for finding food and smelling attackers. The inside of their nose is lined with many smell receptors, which are situated close to their nostrils.

How do you detect taste?

THE SURFACE OF THE TONGUE HAS ABOUT 10,000 MICROSCOPIC TASTE BUDS sunk in it. As you chew, tiny particles of food dissolve in saliva and trickle down to the taste buds. The taste receptors react and send messages about the taste to the brain. There are four basic tastes – sweet, salty, bitter, and sour – and every taste is made up of one or a combination of these. The taste buds in different parts of the tongue react mainly to one of these basic tastes.

How does the sense of touch work?

A taste bud

Taste cells react to chemicals dissolved in saliva.

THE SENSE OF TOUCH TELLS YOU WHETHER SOMETHING IS ROUGH, SHINY, WET, COLD, and many other things. There are many different kinds of sense receptors in the skin, which between them react to touch, heat, cold, and pain. Some touch receptors react to the slightest thing, while others need a lot of pressure to make them respond. The brain puts together all the different messages to tell you how something feels.

Which part of the body is least sensitive to touch?
The back is one of the least sensitive areas of the body.

Does taste matter?
Unpleasant tastes can warn you when food has gone bad or is poisonous. Your body needs healthy food, so enjoying the taste of it encourages you to eat.

How do blind people use touch to see?
Blind people can tell what something is like by feeling it. Outside, they may use a long cane to feel the way in front of them. Blind people read by touch: they run their fingertips over Braille patterns of raised dots that represent different letters.

The tongue

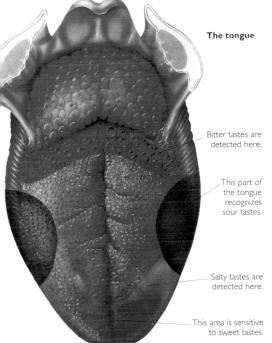

Bitter tastes are detected here.

This part of the tongue recognizes sour tastes.

Salty tastes are detected here.

This area is sensitive to sweet tastes.

Which parts of the body are most sensitive to touch?
Any part of the body that has lots of touch receptors is particularly sensitive to touch. These parts include the lips, tongue, fingertips, and soles of the feet.

Which parts of the body are most sensitive to heat?
Your elbows and feet are more sensitive to heat than many other parts of the body. You may have noticed that bathwater feels much hotter to your feet than it does to your hand. Your lips and mouth are very sensitive to heat too.

Why do you like some tastes better than others?
Most people prefer things that taste sweet or slightly salty, but your sense of taste can easily become used to too much sugar and salt. How you like food to taste is very much decided by your eating habits.

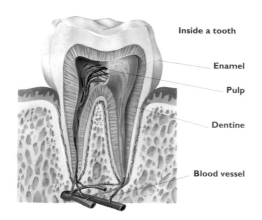

Inside a tooth

— **Enamel**

— **Pulp**

— **Dentine**

— **Blood vessel**

What are teeth made of?

The outside of a tooth is made of enamel and is the hardest substance in the body. Underneath is dentine, which is as hard as bone, and in the center of each tooth is a tender, pulpy mass of blood and nerves.

How many teeth do you have?

Each person has two sets of natural teeth during their life. The first set of 20 are called milk teeth and start to appear at about the age of six months. From about the age of six years, the milk teeth are gradually replaced by 32 adult teeth.

Why are teeth different shapes?

DIFFERENT TEETH DO DIFFERENT JOBS TO HELP YOU CHEW UP FOOD. THE BROAD, flat teeth at the front, slice through food when you take a bite. They are called incisors. The pointed canine teeth are like fangs, and grip and tear chewy food such as meat. The large, flat-topped premolars and molars grind the food between them into small pieces, which mix with saliva to make a mushy ball, ready for swallowing.

Why does vomit taste sour?

When you vomit you bring back partly digested food into your mouth. It tastes sour because it is mixed with acid made by the lining of the stomach. The acid kills germs and helps to break the food into smaller pieces.

Fruit, vegetables, and whole-meal bread contain plenty of fiber. Fiber makes the muscles in the intestines work better.

Healthy food

Which foods give you energy?

Foods such as bread, rice, potatoes, and spaghetti contain a lot of carbohydrates. Carbohydrates give you energy to move, work, and grow. Fats and sugars also give you energy.

Which foods make you grow?

Foods such as milk, cheese, fish, meat, and beans contain a lot of protein, a substance that the body needs to make new cells. A varied and balanced diet contains all the different substances that the body needs to grow.

Where does food go after it is swallowed?

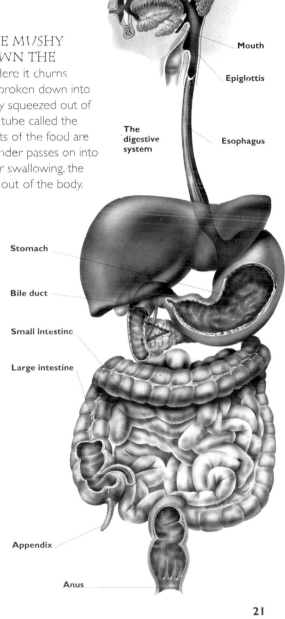

WHEN YOU SWALLOW, THE MUSHY BALL OF FOOD GOES DOWN THE gullet or esophagus into the stomach. Here it churns around for up to four hours, while it is broken down into chyme, a soupy liquid. It is then gradually squeezed out of the stomach and through a long, coiled tube called the small intestine. Here the nourishing parts of the food are absorbed into the blood and the remainder passes on into the large intestine. About 24 hours after swallowing, the remaining waste, called feces, is pushed out of the body.

Mouth

Epiglottis

The digestive system

Esophagus

Stomach

Bile duct

Small intestine

Large intestine

Appendix

Anus

What is the epiglottis?
The epiglottis is a kind of trapdoor that closes off your windpipe when you swallow. It stops food going down into the lungs.

How big is your stomach?
An adult's stomach holds about 1³/₄ pints (1 liter) of food, and a child's a bit less. Your stomach gets bigger the more you eat. A large adult can eat and drink up to 7 pints (4 liters) of food and liquid at one meal.

How long are the intestines?
The small intestine is more than three times as long as the whole body! In an adult this is about 21 ft (6.5 meters). The large intestine is a further 5 ft (1.5 meters) and the whole tube from mouth to anus measures about 30 ft (9 meters).

What is the appendix?
The appendix is a spare part of the large intestine that plays no part in digestion. Sometimes the appendix becomes infected and has to be removed.

Why do feces smell?
Bacteria in the large intestine help to break down waste material, but they also make it smell.

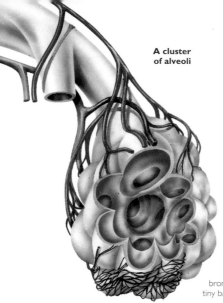

A cluster of alveoli

Why do the lungs have so many alveoli?

In order to provide a huge surface across which oxygen and carbon dioxide can move in and out of the blood. In fact the lungs have over 700 million alveoli. If an adult's alveoli were all laid out flat, they would cover seven car-parking spaces!

Why does your breath sometimes look misty?

The air you breathe out contains water vapor. On a cold day this condenses into a mist of tiny water droplets.

At the end of each bronchiole is a cluster of tiny balloons called alveoli.

The bronchial tubes divide into tiny tubes called bronchioles.

How long can you hold your breath?

You can probably hold your breath for about a minute. The longer you hold your breath the higher the carbon dioxide level in your blood rises, and the more you feel the need to breathe out.

The windpipe is attached to loops of cartilage to make sure the airways stay open.

The lungs

How do you breathe in and out?

The lungs do not have their own muscles to make you breathe in and out. The muscles between your ribs and the diaphragm – a sheet of muscle under the lungs – do the job instead. As your ribs move up and out, the diaphragm contracts and moves down. This pulls air into the lungs to fill the space. When the diaphragm and the muscles between the ribs relax, they squeeze air out of the lungs.

Breathing in and out

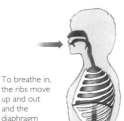
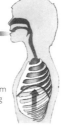

To breathe in, the ribs move up and out and the diaphragm moves down.

To breathe out, the ribs and diaphragm relax, pushing air out of the lungs.

When the diaphragm moves down, it produces space in the lungs, which is filled by taking in air.

The diaphragm is attached to the lungs.

How much air do your lungs hold?

An adult's lungs hold about 10 pints (6 liters) of air, while a child's lungs hold less. You usually breathe about 16 to 20 times a minute, taking in less than 1 pint (0.5 liters) of air each time.

Why do you need to breathe in air?

THE AIR CONTAINS A GAS CALLED OXYGEN, WHICH THE BODY NEEDS TO STAY ALIVE.

When you breathe in, you pull air through your mouth or nose into the windpipe and through narrower and narrower tubes in the lungs. At the end of each tiny tube, or bronchiole, are hundreds of minute balloons called alveoli. As these balloons fill with air, oxygen passes from them into the blood vessels that surround them. The blood then carries the oxygen to all parts of the body. At the same time, waste carbon dioxide passes out of the blood and into the lungs. It leaves the body in the air you breathe out.

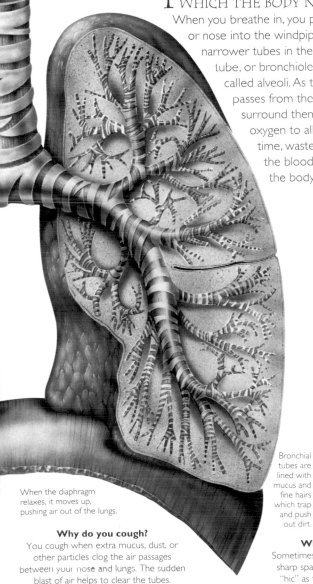

When the diaphragm relaxes, it moves up, pushing air out of the lungs.

Bronchial tubes are lined with mucus and fine hairs which trap and push out dirt.

How do you talk?

When you breathe out, the air passes over the vocal cords in the voice box or larynx in the neck. When the cords vibrate they make a sound. Changing the shape of your lips and tongue makes different sounds, which can be put together into words.

What happens when you sneeze?

When you sneeze, air rushes down your nose like a minihurricane at 100 mph (160 km/h) — up to 20 times faster than normal. Sneezing blasts away whatever dust, pollen, or germs may be irritating your nose.

Why does running make you puff?

Muscles use up oxygen as they work. When you run, your muscles are working hard and need extra oxygen. Puffing makes you breathe in up to 20 times more air, to supply your muscles with the oxygen they need.

Why do you cough?

You cough when extra mucus, dust, or other particles clog the air passages between your nose and lungs. The sudden blast of air helps to clear the tubes.

What happens when you hiccup?

Sometimes the diaphragm begins to contract in short, sharp spasms. These sudden movements make you "hic" as the gulps of air pass over the vocal cords.

23

How often does the heart beat?
A child's heart usually beats about 80 times a minute, a bit faster than an adult's (70 times a minute). When you run or do something strenuous, your heart beats faster to send more blood to the muscles.

How big is the heart?
The heart is about the same size as your clenched fist. It lies nearly in the middle of your chest and the lower end tilts toward the left side of the body.

What is the heart made of?
The heart is made of a special kind of muscle, called cardiac (heart) muscle, which never gets tired.

What do white blood cells do?
White blood cells surround and destroy germs and other intruders that get into the blood.

What is plasma?
Just over half the blood is a yellowish liquid called plasma. It is mainly water with molecules of digested food and essential salts dissolved in it.

How much blood do you have?
An average man has 9 to 10 pints (5 to 6 liters) of blood; an average woman has about 8 pints (4 to 5 liters). Children have less depending on how tall and heavy they are.

The heart

"Used" blood enters the heart from the arms and head.

"Used" blood leaves the heart for the right lung along the right pulmonary artery.

In this diagram arteries are red and veins are blue. Blood stocked with oxygen is red and "used" blood is blue.

Blood stocked with oxygen enters the heart from the right lung along the right pulmonary veins.

The right side of the heart takes in "used" blood and pumps it to the lungs.

What job does your heart do?

THE HEART'S JOB IS TO PUMP BLOOD TO THE LUNGS AND THEN ALL AROUND THE BODY. The right side of the heart takes in blood from the body and pumps it to the lungs. The left side of the heart takes blood filled with oxygen from the lungs and pumps it to the rest of the body. Valves inside the heart stop blood flowing the wrong way. You can feel your heart beat if you put your hand on your chest.

"Used" blood from the lower body returns to the heart.

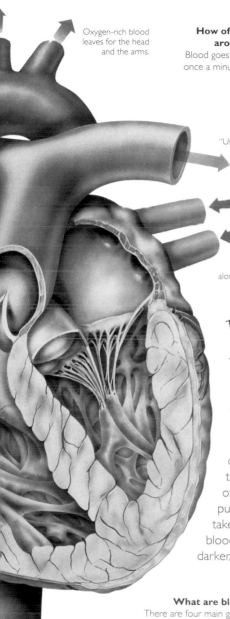

Oxygen-rich blood leaves for the head and the arms.

How often does blood go around the body?
Blood goes around the body about once a minute or 1,500 times a day.

What is a blood transfusion?
If a person loses a lot of blood, perhaps due to an accident or operation, the lost blood can be replaced with blood given by someone else. The new blood is dripped straight into a vein.

"Used" blood leaves the heart to go to the left lung along the left pulmonary artery.

What is a capillary?
Blood travels around the body through tubes called arteries and veins. These branch off into smaller and smaller tubes that reach every cell of the body. Capillaries are the tiniest blood vessels of all. Most capillaries are thinner than a single hair. If an adult's capillaries were laid end to end they would stretch 60,000 miles (100,000 km) — nearly 2½ times around the world.

Blood stocked with oxygen comes to the heart from the left lung along the left pulmonary veins.

Why is your blood red?

EACH TINY DROP OF YOUR BLOOD CONTAINS UP TO 5 MILLION RED blood cells that give blood its color. Red blood cells contain a substance called hemoglobin, which takes in oxygen in the lungs. Blood that is rich in oxygen is bright red. As this bright red blood is pumped around the body, the oxygen is gradually taken up by the body's cells. By the time the blood returns to the heart, it is a slightly darker, more rusty red.

Blood in close-up

Plasma

What are blood groups?
There are four main groups of blood, called groups A, B, AB, and O. Only some groups can be mixed with others, so doctors find out which blood group a patient belongs to before giving a blood transfusion.

Red blood cell

Oxygen-rich blood leaves the heart to be pumped around the body.

White blood cell

What is bile?

Bile is a yellow-green liquid made by the liver and stored in the gallbladder. From there it passes into the small intestine, where it helps to break up fatty food.

Why is urine yellow?

Urine contains traces of waste bile and this makes it yellowish. If you drink a lot of water, your urine will be diluted and less yellow, but the first urine of the morning is usually stronger and darker. Some foods affect the color of urine. Eating beets can turn it pinkish.

Kidneys and bladder

A large artery called the aorta brings blood from the heart to the kidneys.

Blood is filtered in the kidneys and the waste urine is funneled into the ureters.

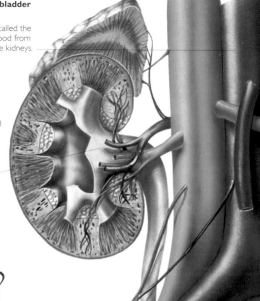

What does the liver do?

THE LIVER IS A CHEMICAL FACTORY THAT DOES MORE THAN 500 DIFFERENT JOBS. Some of its most important functions concern the processing of digested food and the removal of waste and poisons from the blood. Digested food is taken straight from the intestines to the liver, where some nutrients may be released into the blood and the rest stored to be used later. The liver processes poisons, and changes unwanted proteins into urea. The kidneys remove poisons and urea and make them into urine.

What makes you urinate?

The bladder stretches as it fills. When it has about ¼ pint (150 ml) in it, nerves in the walls of the bladder send signals to the brain and you feel the need to urinate.

As urine trickles down the ureters and into the bladder, the bladder stretches.

How much urine does the bladder hold?

An adult's bladder can hold up to about 1 pint (600 ml) of urine, and a child's holds less. But you usually need to go to the bathroom as soon as your bladder is about a quarter full.

Cleansed blood leaves the kidneys and returns to the heart.

Urine leaves the bladder through the urethra when a circle of muscle relaxes to open the entrance to the tube.

The kidneys nearly 2 pints (1 liter) of blood every minute.

The ureter is a long tube that takes urine from the kidneys to the bladder.

Muscles in the ureter help to move the urine through the tube.

What do the kidneys do?

KIDNEYS FILTER THE BLOOD TO REMOVE WASTES AND EXTRA WATER AND SALTS. Each kidney has about a million tiny filters, which between them clean about a quarter of your blood every minute. The kidneys work by forcing many substances out of the blood and then taking back in only what the body needs. The unwanted substances combine with water to make urine, which trickles down to the bladder where it is stored.

How much liquid do you need to drink?
You need to drink about 2 to 2 ½ pints (1.2 to 1.5 liters) of watery drinks every day. The amount of water you take in balances the amount you lose. Most water is lost in urine and feces. But sweat and the air you breathe out also contain water.

How long can you live without water?
Although some people have lived for several weeks without food, you can survive only a few days without drinking water.

Why do you need to drink more in hot weather?
When it is hot, you sweat more and so lose more water, which you then replace by drinking more.

Why do you sweat when you are hot?
Sweating helps to cool you down. When the body becomes hot, sweat glands pump lots of salty water onto the skin. As the sweat evaporates (changes into water vapor), it takes extra heat from the body.

Why do you go red when you are hot?
As different parts of the body burn up energy they make heat. Blood carries the heat around the body. If the body becomes too hot, the tiny blood vessels near the surface of the skin expand to help the blood cool. Blood flowing near to the skin makes the skin look red.

Athlete sprinting

When muscles work hard, they produce heat as well as movement.

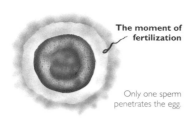

The moment of fertilization

Only one sperm penetrates the egg.

What do babies do in the womb?

As the unborn baby gets bigger, it exercises its muscles by kicking, moving, and punching. It also sucks its thumb sometimes, opens and shuts its eyes, and goes to sleep.

What is labor?

Labor is the process of giving birth. The neck of the womb stretches and opens, and then the womb, which is made of strong muscle, contracts to push the baby out. Labor can take several hours.

What are genes?

Genes are a combination of chemicals contained in each cell. They come from your mother and father and determine everything about you, including the color of your hair, how tall you will be, and even what diseases you might get in later life.

Why do children look like their parents?

You inherit a mixture of genes from your parents, so in some ways you will look similar to your mother and in others to your father.

What is a fetus?

A fetus is an unborn baby from eight weeks after conception until birth. In the first seven weeks after conception it is called an embryo. By 14 weeks the fetus is fully formed, but it is too small and frail to survive outside the womb. Babies of 24 weeks can survive in an incubator if they are born early, but most stay in the womb for the full 36 weeks.

What is a condom?

A condom is a thin, rubber sheath that fits over the man's penis. It stops sperm getting into the woman's vagina during sex, so that a baby cannot be conceived. Other contraceptives (things that prevent pregnancy) include the Pill, which is taken by women.

Where does a man's sperm come from?

Sperm are made in the testicles, two sacs that hang to either side of the penis. After puberty the testicles make millions of sperm every day. Any sperm that are not ejaculated are absorbed back into the blood.

Male reproductive organs

What is a period?

If the egg is not fertilized by a sperm, it passes out of the woman's body through the vagina. At the same time, the lining of the womb and some blood also pass out of the body. This slow flow of blood lasts about five days every month and is called a period.

Where does the egg come from?

When a girl is born she already has thousands of eggs stored in her two ovaries. After puberty, one of these eggs is released every month and travels down the Fallopian tube to the womb.

Female reproductive organs

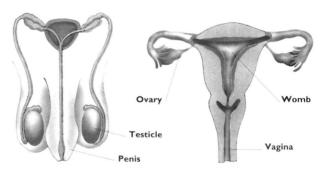

Ovary

Testicle

Penis

Womb

Vagina

Testicles hang outside the body to keep the sperm cool.

The vagina joins the womb to the outside of the body.

How does a new baby begin?

A NEW BABY BEGINS WHEN A SPERM FROM A MAN JOINS WITH AN EGG from a woman. This is called fertilization, and it happens after the man ejaculates sperm into the woman's vagina during sex. The cells of the fertilized egg begin to multiply into a cluster of cells, which embeds itself in the lining of the womb. There the cells continue to multiply and form the embryo of a new human being.

How fast does an unborn baby grow?

YOU GROW FASTER BEFORE YOU ARE BORN THAN AT ANY OTHER TIME IN YOUR LIFE. Three weeks after the egg is fertilized, the embryo is no bigger than a grain of rice. Five weeks later, almost every part of the new baby has formed – the head, brain, eyes, heart, stomach, and even the fingers – yet it is only about the size of a thumb. By the time it is born, 30 weeks later, it will probably be about 20 in (50 cm) long and weigh about 7 lb (3.5 kg).

How does an unborn baby feed?
Most of the cluster of cells that embeds itself in the womb grows into an organ called the placenta. Food and oxygen from the mother's blood pass through the placenta into the blood of the growing baby.

A baby in the womb

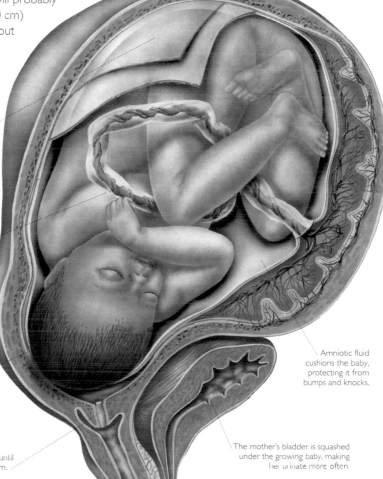

The placenta supplies the unborn baby with oxygen and food from the mother's blood.

The umbilical cord joins the baby to the placenta.

This baby is ready to be born headfirst.

Amniotic fluid cushions the baby, protecting it from bumps and knocks.

The cervix stays tight shut until the baby is ready to be born.

The mother's bladder is squashed under the growing baby, making her urinate more often.

Why do babies cry?
A baby cries when it needs something – when it is hungry or lonely. It also cries when it has a stomachache or other pain.

Are newborn babies completely helpless?
A baby can breathe, suck, and swallow from the moment it is born, so it is not completely helpless.

When are you fully grown?
Boys and girls grow quickly during puberty, and then they grow more slowly until they reach their full height some time around age 20.

What do all new babies need ?

Aᴸᴸ NEWBORN BABIES NEED FOOD, WARMTH, LOVE, AND PROTECTION. At first a baby can only drink liquids, so it sucks milk from its mother's breasts or from a bottle. Milk contains everything the new baby needs to grow and stay healthy. A baby also needs to be washed and have its diaper changed regularly. Babies sleep a lot of the time, but when they are awake they need plenty of smiles and cuddles. Babies and children rely on their parents for the things they need.

What happens when a boy's voice breaks?
A boy may be growing so fast during puberty that the muscles that control his vocal cords cannot keep up. His voice may suddenly change from high to low before finding the right pitch. The vocal cords also become thicker, making his voice deeper.

Stages of growth

At puberty the sexual organs begin to mature.

During childhood, the legs and arms grow longer and the child becomes more adept and confident.

A two-year-old is about half the height it will be when adult.

When do babies learn to walk and talk?
By its first birthday, a baby is usually already pulling itself up on to its feet and is nearly ready to walk. It may also be beginning to say a few words, though talking develops slowly over the next few years.

Why do you have to support a young baby's head?
When a baby is born the muscles in its back and neck are very weak, too weak to hold up its own head. A baby's head is much bigger for the size of its body than a child's or an adult's.

Why can't young babies sit up?
Young babies cannot sit up until their back muscles have grown strong enough to support them. This happens around 6 months.

Babies often learn to crawl before they take their first tottering steps.

What are hormones?
Hormones are chemicals released into the blood from various glands. Some glands make sex hormones that control the menstrual cycle.

Adults are fully grown and may decide to have children of their own.

What is puberty?

PUBERTY IS THE TIME IN WHICH YOU GROW FROM A CHILD INTO AN ADULT. You grow taller and your body changes shape. A girl develops breasts and her hips become broader. Her waist looks thinner. A boy's chest becomes broader and his voice grows deeper. At the same time, the sex organs develop. A girl begins to have periods and a boy begins to produce sperm. Puberty lasts several years and affects moods, feelings, and attitudes as well as bringing physical changes.

As people get older, they begin to slow down.

Very old people may become quite frail.

Why do people age?
The cells of the body are constantly being renewed, except for brain cells and other nerve cells. As people get older, the new cells do not perform as well as the cells of younger people.

How long do most people live?
Most people in the developed world live until they are over 70 and more and more people are living into their 80s and beyond.

What makes you grow?
A growth hormone tells your body to grow. This is produced in the pituitary gland in the brain and taken all around the body in the blood. Exactly how tall you grow is determined by genes inherited from your parents.

What is the menopause?
The menopause is when a woman's body changes so that she is no longer able to have children. As sex hormone levels drop, her ovaries stop producing eggs. The woman may experience uncomfortable, hot flushes and unpredictable mood swings.

Who had the most children?
It is believed that a Russian woman who lived in the 1700s holds this record. She was called Madame Vassilyev and she gave birth to no fewer than 69 children.

Index